Scrum First Aid

A Recipe Book for Fixing Scrum Challenges and Anti-Patterns

Todd A. Jacobs

Scrum First Aid

A Recipe Book for Fixing Scrum Challenges and Anti-Patterns

Todd A. Jacobs

© 2020 Todd A. Jacobs

Tweet This Book!

Please help Todd A. Jacobs by spreading the word about this book on Twitter!

The suggested tweet for this book is:

I just bought #ScrumFirstAid, a book about fixing real-world #Scrum challenges and #antipatterns. Highly recommended! http://leanpub.com/scrum-first-aid

The suggested hashtag for this book is #ScrumFirstAid.

Find out what other people are saying about the book by clicking on this link to search for this hashtag on Twitter:

#ScrumFirstAid

Also By Todd A. Jacobs
The Agile CIO

Contents

Dedication i

Epigraph iii

Introduction v

How to Read This Book vii
    Pick Your Reading Style vii
    Chapter Navigation viii

Anti-Patterns 1

Daily Scrum as Status Pulls 3
    Illustrative Example 3
    Analysis and Advice 5

Sequential Development and Testing 11
    Illustrative Example 11
    Analysis and Advice 12

Expecting Ever-Increasing Velocity 19
    Illustrative Example 19
    Analysis and Advice 20

Sprint Names as "Permalinks" 23
    Illustrative Example 23

Analysis and Advice	23
Incomplete Work as "Failure"	29
Illustrative Example	29
Analysis and Advice	31

Implementation Questions · · · · · 35

How are Product Releases Scheduled?	37
Illustrative Example	38
Analysis and Advice	38
Why Define a Sprint Goal Each Sprint?	43
Illustrative Example	43
Analysis and Advice	43
Who Manages Project Budgets?	49
Illustrative Example	49
Analysis and Advice	50
Must Sprint Backlogs Be Completed?	53
Illustrative Example	53
Analysis and Advice	54
What is Scrum's Process Overhead?	57
Illustrative Example	57
Analysis and Advice	57

Backmatter · · · · · · 63

About the Author	65

Notes · · · · · · 68

Dedication

- **For my father Martin,**
  who never understood what I do for a living, but who was proud of me anyway.

- **For my wife Kim,**
  who supports everything I do.

- **For my son Avi,**
  whose enthusiasm for new things encouraged me to write this book.

Epigraph

Agility is the engine of modern business, and automation is its fuel.

— Todd A. Jacobs

Introduction

The Scrum Guide is only 19 printed pages long, including cover pages, indexes, and *et cetera*. It also omits *any* mention of the roles and responsibilities of line management and senior leadership, focusing instead on a little theory and some essential details about framework events. This brevity often leads people to believe that Scrum is simple, or that implementing it is easy, and that it can be done without being embraced across the entire organization.

In the real world, I've seen non-agile organizations send a single representative of the team to collect a Scrum Master certification, rebrand a few line line managers as "Product Owners," and then tell a bunch of individual contributors that they are now part of a self-sufficient Scrum Team. What could possibly go wrong with that?

In these scenarios, the leadership team often sits back, confident that sprinkling the right agile buzzwords across the annual strategic plan has proactively fixed all the business problems they're facing. Everything will stay the same, only better! Things will move faster; quality will improve; time to market will drop; executive bonuses will roll in!

When adopting Lip-Service Agile™ or employing Buzzword Management℠ fails, the blame is often laid at the feet of:

- a "wrong-headed" (rather than simply misimplemented) framework like Scrum;
- the "team" of random individual contributors who failed to become agile enough, fast enough;
- the Scrum Master who failed to hold the team *accountable* to management targets.

In almost all of these cases, improper application of framework principles, values, and roles is really the foundational problem. Failure to embrace Scrum at the senior leadership level (often called "tone at the top") is a common comorbidity of a dead or dying Scrum implementation. In contrast, successful Scrum requires all hands on deck, all the way from the C-level down to the development team.

This book addresses a cross-section of common Scrum problems. Each chapter starts with a real-world problem an organization has applying Scrum to a concrete problem. Once the underlying problem has been identified, each chapter then suggests process improvements that will get things back on track.

As Fred Brooks famously argues in several publications, there is "no silver bullet." Scrum is not a magic incantation that will fix organizational dysfunction, but it *is* a solid framework based on empirical control principles. This book will show you some concrete ways to apply those framework principles to the very real problems you face, and help teams and organizations struggling with Scrum to fix the *process* problems that are standing in the way of their success.

How to Read This Book

Pick Your Reading Style

The format of this book is like a box of chocolates. You don't have to eat the whole box all at once. While astute readers will find themes and core principles that run through the whole book, each chapter focuses on a single problem or set of circumstances that may apply to your situation.

Flipping through the index to find situations close to yours, and then reviewing the associated recommendations, is a great way to start. It's like having an agile coach in your pocket who can offer you advice on how others have solved similar problems.

Reading the book cover-to-cover is a good option, too. If you read the book this way, look for common themes and principles that you can add to your agile toolbox. Thematic items are called out as tips throughout the book, but a more thorough reading will reward you with a broader agile perspective.

Regardless of how you approach the book, the primary goal is to frame-shift from "Scrum as a set of arbitrary practices" to a process-oriented view that leverages the agile values and principles built into the Scrum framework to continuously improve *your* organization's processes.

There's no silver bullet. However, you're likely to discover that reframing Scrum implementation problems as process improvement opportunities is the next best thing.

Chapter Navigation

This book is split into two main sections to help readers find the material they need most.

1. Anti-Patterns

    Each chapter in this section corresponds to a Scrum anti-pattern where practices are actively working against the success of the Scrum Team and the product development process.

2. Implementation Questions

    The chapters in this section address common questions about how Scrum works, and how to leverage the framework to best advantage.

Within each chapter, you will find a number of tips, ideas, and warnings. These are indicated as follows:

 This is a tip or key insight.

 This is a warning that you should take to heart.

 This is additional information you may find useful.

You will also encounter footnotes, endnotes, and reference links throughout each chapter, formatted appropriately for your ebook reading device. These items are supporting information or clarifications that aren't important enough to disrupt the narrative flow, but that are often interesting enough to click on. Endnotes are also used to cite sources.

Anti-Patterns

This section will address common Scrum anti-patterns. Each chapter corresponds to a Scrum implementation where practices are actively working against the success of the Scrum Team and the product development process.

Daily Scrum as Status Pulls

Illustrative Example

For many teams new to Scrum, the Daily Scrum is a core event that seems straightforward, but often presents numerous implementation problems when transitioning from a more traditional command-and-control process. An almost-universal challenge is differentiating between the Daily Scrum and a status report meeting. Here's a typical example.[1]

I'm a developer, and my team started using Scrum last year. It has been rough to say the least. I was always taught that the Daily Scrum was to discuss the following:

1. What you did yesterday.
2. What you are going to do today.
3. Any issues, challenges, or roadblocks that might be in the way at the moment.

My understanding is that this meeting is for developers, but Product Owners should be there to answer any questions that may be directed toward them. Otherwise, this is a meeting to go over developer needs and issues in a manner that is relaxed, and for the benefit of the developers (no one else).

In our version, each developer is required give what is tantamount to some kind of regimented status report. We stand up and report in the following format:

1. Story number and story name.
2. Each task number in the story that we worked on.
3. The original time estimate for the currently-discussed task.
4. The actual time estimate for the currently-discussed task.
5. An explanation of why we were over or under the original estimate.

We repeat this for each story and task in the current Sprint. Is this Scrum? Heck, is this even agile?

Analysis and Advice

When Daily Scrums Become Status Pulls

What is happening here is *not* an effective Scrum stand-up meeting; it's just a traditional status pull. In fact, it's probably worth dissecting this particular status pull to see why this particular "Scrum" process is failing. Two key elements that the project manager (not Scrum Master, evidently) is asking are:

- The original time estimate for the currently-discussed task.
- The actual time estimate for the currently-discussed task.

The only *practical* utility for this information is to determine whether estimates were on-target or not, or whether there are hidden process impediments that were not planned for in the original estimate. However, gathering this information in this particular way is not very agile because:

1. This information should already be transparent through the use of a Kanban board* or other framework process. Having to ask for it explicitly is a "project smell" that indicates a fundamentally broken project process or an undocumented objective.
2. Estimates are estimates, not commitments. As long as all stories that the team has voluntarily committed to perform during the Sprint are completed by the end of the Sprint, the individual story estimates (as opposed to the aggregate estimate) are not intrinsically useful except as part of a retrospective.

Kanban is a Japanese loanword that literally means "signboard." While the Scrum framework doesn't require the use of a kanban, its use has been widely adopted within the agile community because of its effectiveness as an "information radiator."

3. If the process is based on 100% utilization, rather than a throughput-based pull queue, then it's neither Scrum nor agile. All agile frameworks require slack in the process; asking for daily deltas at such a granular level certainly implies an intolerance for the slack *required* by agile frameworks.

Now, the last item is particularly telling. The project manager wants:

- An explanation of why we were over or under the original estimate.

Getting better at estimating is a useful goal for any agile framework. However, one goal of Scrum is to ensure that the team does not overcommit; if there is extra capacity within the Sprint once all stories are completed, then the team can and should pull additional stories into the Sprint from the Product Backlog.

The question, as described, sounds a lot like stories are being *assigned* to the team for each Sprint, which is an epic fail from a Scrum standpoint. Even if that's not actually the case, it's another project smell* that indicates that individuals are being "held accountable" (presumably by the PM) for the accuracy of individual story estimates, rather than for the team's overall progress in meeting the Sprint Goal for that iteration.

Misestimating is an issue that should always be communicated clearly during Sprint Reviews, and used as a learning process during the Sprint Retrospective. However, the *tone* of the question implies that accountability for estimates is more important than the work itself, and diverts the Scrum Team's focus from feature delivery to "C.Y.A. delivery."

---

*In agile software development, a code smell is an indicator that there may be a non-obvious problem that merits investigation. While a whiffy smell doesn't *prove* that there's a problem, it's an early warning that the team should take a closer look. It's such a useful metaphor that this book often applies the term "project smell" to indicate a process problem or framework antipattern that should be carefully explored and re-evaluated.

Using the Stand-Up to Coordinate Status

The following is partially correct, but misses the essence of what the
Daily Scrum is for.

> [The Daily Scrum] is a meeting to go over developer needs and issues in a manner that is relaxed, and for the benefit of the developers (no one else).

The Daily Scrum is a commitment and coordination meeting between members of the Development Team, but as an information radiator it can benefit the entire team. It's designed to ensure that the entire team is aware of impediments, what stories are done or not-done, and what tasks are ready to be pulled from one team member's to-do list into someone else's.

Despite being a meeting primarily for the developers, a Scrum Master or Product Owner can also gain value from the meeting. A well-run daily stand-up will give the Scrum Master a clear picture of any process issues that need attention, and whether individual stories are done or not-done. The Product Owner gains a sense of whether there are risks to the current Sprint Goal, timely insight into Backlog Refinement tasks for the next Sprint, and advance notice of when a Sprint might need to be terminated early.

It is often important that the Scrum Master and the Product Owner be present at the Daily Scrum, but primarily as *passive* participants. However, if the team is *reporting* to either of them, then the team is neither self-organized nor empowered to efficiently deliver value. Status-reporting to an authority figure during the Daily Scrum is definitely a framework anti-pattern, and one that deserves immediate attention.

The Scrum Master's role in the Daily Scrum is primarily to act as a process referee when necessary, although a little coaching and meeting facilitation may come with the territory during early phases of Scrum adoption.

The Product Owner's role in the meeting is as an on-demand resource, to clarify any questions the Development Team might have about the scope or intent of a Product Backlog item.

Your Process Might Be Broken If...

Here is a short list of "project smells" related to status-reporting issues. If any seem applicable, address them at the next Sprint Retrospective.

1. If the team has so many stories on the board at once that a verbal report needs to identify the story by number, then the process is broken.
2. If the team has so many stories that a glance at the board doesn't make it clear that things are moving from not-done to done in a timely manner, then the process is broken.
3. If the team is so large that a glance at the board doesn't indicate who is working on what—or worse yet, requires cross-referencing with a separate spreadsheet—then the process is broken.
4. If the Product Owner's only communication about the project or with the team is at the daily stand-up, then the process is broken.
5. If the Scrum Master can't construct a burn-down chart from some combination of the story board, the daily stand-up, and ongoing communication with team members, then the process is broken.

6. If the Daily Scrum is not serving the needs of the team, then the process is broken.
7. If estimates encourage excuses or justifications rather than improved Sprint Planning techniques, then the process is broken.
8. If the team is reporting to any one individual during the Daily Scrum rather than cooperatively coordinating with one another, then the process is broken.
9. If your process is broken, but team members don't feel free to directly address the broken process during a Sprint Retrospective, then your organizational culture is broken.
10. If your organizational culture is broken, and figuring out *Who do we blame for breaking it?* is more important than figuring out *How do we collectively fix it?*, then it's time to dust off your resume.

Sequential Development and Testing

Illustrative Example

Sometimes, teams transitioning to Scrum have a lot of implicit assumptions about product or software development should work. Even experienced agile teams can occasionally fall into the trap of allowing unexamined assumptions to throw sand in the gears of the development process.

The following example[2] shows how unexamined assumptions about collective ownership and process hand-offs can negatively impact a Scrum implementation.

What are developers expected to do during testing in the latter half of each Sprint? When you are using the Scrum framework, a Sprint cycle involves both development and testing. At the end of the Sprint, the tasks worked upon and tested during the iteration are showcased and released.*

Typically, a team of three to four developers would have one quality assurance (QA) resource to perform testing.† What are the developers expected to do when testing is happening? Since the number of developers is much higher than the number of testers, the developers often fix bugs found by the testers very quickly, leaving the developers with nothing to do‡ towards the end of the Sprint.

When practicing Scrum, what is expected of the developers during QA testing?

Analysis and Advice

The question embeds some false assumptions about the linear nature of development and testing within an agile process. A mature agile team with cross-functional skills should generally treat development and testing as intertwined activities, not as sequential ones.

---

*The apparent disconnect between continuous delivery and providing a "potentially shippable increment" each Sprint is a a red herring. While out of scope for this chapter, always watch out for this implicit assumption!

†This is an implicit segmentation into sub-teams, which is explicitly prohibited by the Scrum framework.

‡This is an implicit appeal to the "100% utilization fallacy," which posits that optimum results come from keeping everyone on the team as busy as possible. While possibly true in some domains, agile frameworks like Scrum, Lean, and Kanban *require* a certain amount of slack in the process. As explained by queueing theory, insufficient slack typically *lowers* productive throughput.

Mature Scrum Teams integrate development and testing so that they are not fundamentally separate work streams. Failing that, the Scrum Team must formally accept the risks and process deficits associated with a stovepiped workflow.

"Teams" That Aren't Cohesive, Cross-Functional Teams

One of the most critical implicit assumptions in the question is that quality assurance activities and testers are somehow separate from the rest of the Scrum Team. The Scrum Guide's section describing the Development Team provides some reasons why this is an anti-pattern.

- Scrum recognizes no titles for Development Team members, regardless of the work being performed by the person;
- Scrum recognizes no sub-teams in the Development Team, regardless of domains that need to be addressed like testing, architecture, operations, or business analysis; and,
- Individual Development Team members may have specialized skills and areas of focus, but accountability belongs to the Development Team as a whole.

QA Must be Embedded, Not a Separate Process Track

Since the number of developers is much higher than the number of QA testers, when bug fixes get done very quickly the developers are left with nothing to do towards the end of the Sprint.

The workflow described here is *not* Scrum. It's much more like typical waterfall development. Scrum activities should be as cross-functional as possible, and leverage a multifaceted team-based approach to all tasks.

As one example, testers and developers should be working in lock-step throughout a Sprint. Testers should be involved early and often, helping the developers design testable features by working on test criteria *from the beginning* before a single line of code is written, and helping to ensure that tests are written first.

The Test-First Mindset

Test-Driven Development (TDD) and Behavior-Driven Development (BDD) are most often understood as technical practices. As an empirical control process, Scrum uses the Definition of Done as a way to integrate a test-first approach into the framework. When combined with empirical control, just-in-time planning, and emergent design principles, the framework clearly intends for integrated testing to be a core principle.

The whole team should be running continuous integration tests every single day, so that there's a tight (and ideally immediate) feedback loop between development and testing. By working *with* the developers, rather than being treated as a separate follow-on activity, QA becomes an intrinsic part of the design and development cycles rather than an externality.

Developers and QA Should Partner for Testing

What are the developers expected to do when QA is happening. Since the number of developers is much more than the QA, bug fixes get done very quick and developers are left with nothing to do towards the end of the sprint.

Just as testers are expected to be involved with the developers from day one, developers are expected to work with QA during testing tasks. Rather than tossing code over the wall to testers, developers and testers should work together on the testing process so that bugs are fixed as they are discovered.

Imagine a pair-programming scenario where a tester and a developer work together on a test suite. Instead of a developer dumping a wall of code on the tester and then waiting for results, the two might work *together* on tests and refactorings. For example, compare this highly-collaborative conversation to more traditional, ticket-based handoffs:

QA
The X widget failed the embiggening test.

DEV
Oops. Okay, I'll refactor the embiggener class while you test the end-user insult generator.

QA
Will do. Oh, look, the insult generator passed!

DEV
Great! While you were working on that, I widened the embiggener. Try it again.

QA
It's embiggening properly now. Let's move on to the next set of specs together!

When All Else Fails, Shine a Light on Dysfunctional Process

If for some reason your team can't or won't cooperatively swarm over test-related activities, then the process must make that cost *visible* to the team. If developers and QA insist on playing volleyball with tasks by tossing things back and forth over a net instead of integrating to perform the work together, then you simply make that (potentially dysfunctional) process fully visible.

Do testers *really* have nothing to do during the first phase of a Sprint? No, but if that's your process then you acknowledge that having testers idling on the team for half a Sprint is one of the costs of doing business. Likewise, if developers really have nothing at all to contribute to the testing process in the latter half of a Sprint, then you explicitly acknowledge that your developers are getting paid to hang out on Facebook for 50% of the time, and can accept that as a cost of doing business within your chosen process.

Healthy teams treat all members as cross-functional resources, with value to add during each step of the process. Even if you choose not to fully integrate testing as a first-class activity within your Sprints, testers and developers can take turns assisting each other on current tasks. For example, during development a tester can work collaboratively to design test fixtures while the developer is writing the feature; then during testing, the developer might run code coverage analysis or work on converting test results into documentation while the tester runs the tests.

If the team can't or won't work cooperatively in this way, then the organization can simply accept the fact that some roles within the team will be idle at certain points in the process. While not ideal, it might be politically necessary to simply acknowledge that 50% of your roles will be idle at any given time, and that this is an acceptable cost of doing business within your current development process. While I personally consider this a sub-optimal option, it is still better than falling prey to the *100% utilization fallacy* that tries to keep everyone looking busy even when doing so is wasteful and generates no value...and sometimes may even actively *reduce* productivity.

Expecting Ever-Increasing Velocity

Velocity is a measure of a team's historical capacity, rather than a valid measure of the team's speed or productivity. It should be used primarily during Sprint Planning to forecast Sprint Backlog capacity for the upcoming Sprint, although it can also be used as a detective control to indicate hidden process problems or unexpected technical debt.

 The goal of Scrum is to have a sustainable and predictable development cadence. As such, it is generally better to aim for consistent velocity over time rather than for ever-increasing velocity.

Illustrative Example

Because velocity is so often misunderstood or misused, it leads many organizations to question why their velocity isn't always trending upwards[3], or to ask why management-defined velocity targets aren't routinely being met.

Some Scrum Masters are concerned with "bumping up" a team's velocity during a Sprint, as if a higher velocity is better. Based on my own understanding, points are a relative measure of the work that a team needs to do to produce feature X; the points aren't high or low, they're just more or less than some other feature the team arbitrarily chose as a point of reference. What's important is that you know what your velocity is, however many points that may represent.

If points represent effort, that effort doesn't always increase over time. As the team gains experience, they may apply the same level of effort, but the work they do ends up being more productive. So, a story that was estimated at five points when the team first formed could eventually be estimated as three or two points as the team increases its skills.

Is that correct? I don't see a consensus on whether points measure complexity or effort, but either way it seems to me that as time passes, the perceived effort or complexity of a task will go down and you will be able to fit more of them in a Sprint, thus keeping your velocity constant.

Analysis and Advice

The simplistic answer is that a project's velocity should only increase until the team has developed **a stable, predictable cadence** that can be maintained over time. There are a few caveats, of course, but it's a solid rule of thumb.

Targeting an indefinite upward trend on velocity is a "project smell" that the velocity metric is being misused or misunderstood. Expecting such a trend is often a symptom of management's desire to press a "go faster" button without respecting the sustainable-capacity limits of a project or team.

Velocity Measures/Forecasts Capacity

Velocity is a measure of a team's historical capacity, rather than a directly-correlated measure of the team's speed or productivity. It should be used primarily during Sprint Planning to forecast Sprint Backlog capacity for the upcoming Sprint.

A team's capacity can certainly change over time: capacity can go down during holiday periods or reorganizations, or go up as teams gain domain knowledge, leverage iterative improvements to their test-driven designs, or pay down technical debt.

However, it's a fallacy to assume that capacity can or should have an upward trend. Even though the cone of uncertainty narrows later in a project, the cost of technical debt and refactorings generally increase, and these factors consume team capacity as well. In addition, it is not uncommon for the *complexity* of user stories to increase over time as the low-hanging fruit is picked from the Product Backlog. Again, this added complexity will consume a commensurate amount of capacity from the team.

The goal of Scrum (and iterative agile development in general) is to have a sustainable and predictable development cadence. As such, it is generally better to aim for *consistent* velocity over time rather than for *increasing* velocity.

Sprint Names as "Permalinks"

Illustrative Example

This is an extremely common question when adopting Scrum. Whether it stems from a tradition of named work-packages, management-defined goals and milestones, or simply a very human desire to *name* things, the question of whether Sprints should have names—and if so, how to name them—is arguably one of the very first questions every new Scrum project needs to address.

The following example[4] is an evergreen question asked by almost every organization that adopts Scrum.

I was wondering if it's a good idea to name Sprints with user-friendly names instead of using numerals like Sprint 1, Sprint 2, etc. Are there best practices or standards for naming the Sprint after themes or major milestones for the project?

Analysis and Advice

As a rule, giving Sprints non-sequential numerical names is an anti-pattern. Don't do it!

A Sprint is a container where Product Backlog items are temporarily stored for a brief duration. A Sprint may produce some project artifacts, but the Sprint itself has value only as a time-box. Even if you can find a use case for tracking Sprints by name, giving Sprints non-ordinal labels reduces ease of communication about the project and limits the value of information conveyed by the name.

Sprints Are Ephemeral

A Sprint is the name the Scrum framework gives to a time-boxed iteration. While each Sprint has a defined Sprint Goal, the Sprint itself is not a project artifact that needs to be kept for historical records or process improvement.

The Sprint Goal is one of the most essential (but often overlooked) aspects of Sprint Planning. Whether a Sprint is successful or not isn't based on how much work is done or effort expended, but rather on whether the Scrum Team has successfully met the current Sprint Goal.

As the Development Team works, it keeps the Sprint Goal in mind. In order to satisfy the Sprint Goal, it implements functionality and technology. If the work turns out to be different than the Development Team expected, they collaborate with the Product Owner to negotiate the scope of Sprint Backlog within the Sprint.

Sprints often result in various outputs that are relevant to the project such as velocity measurements, finished features, new user stories for the Product Backlog, or process improvements, but the time-box itself has no lasting value once it has expired. Regardless of whether the Sprint Goal was met or not, the old time-box is gone forever and a new time-box for the next Sprint replaces it.

Giving code names or handles to a sprint therefore makes little sense, since any given Sprint should have no utility as an historical referent. Trying to map ephemeral time-boxes onto named events is usually a project smell that the development model is not truly iterative.

Sprints Aren't Milestones

While each Sprint has a Sprint Goal, and a set of items peeled off the Product Backlog to meet that goal, neither of these things is necessarily a project milestone. Sprint Goals are not always unique, and stories are sometimes placed back onto the Product Backlog (modified or unmodified) for future Sprints. Therefore, there is no guarantee that a given Sprint is actually unique.

For example, if your current Sprint does not meet its Sprint Goal, the Product Owner could decide to have another Sprint with the same goal, although the Scrum Team may prioritize and select different Product Backlog items to better meet that goal. Would you name both Sprints the same if both the Sprint Goal and the Sprint Backlog items were the same? What if the Sprint Goal was the same, but the contents of the Sprint Backlog were different? In any case, would "Sprint Fuschia" following "Sprint Aquamarine" really communicate anything useful about the Sprint, the project, or the team?

In the database world, the problem of (potentially) non-unique rows is generally solved by assigning auto-incrementing primary keys to each row. This is one reason that Scrum generally uses incremented integers rather than code names for labeling sprints.

Projects Evolve; Sprint Labels Shouldn't

In Scrum, the guiding principle is "inspect and adapt." While a project may have a schedule with defined shipping dates, the contents of any given Sprint is expected to change over the course of the project. Given that the output of one Sprint impacts the contents of subsequent Sprints, it would make no sense to assign meaningful code names to each Sprint ahead of time since the goal or user stories for the 23rd sSprint may change by the time the team gets the first 22 Sprints out of the way.

Alternatively, if you use colorful but meaningless names, what value have you added? At least sequential names give you sequence information. If you name your Sprints after trees, endangered mammals, or exoplanets, what useful information does that actually convey to anyone?

Since we've already established that Sprints are ephemeral and carry no historical significance (other than perhaps as velocity data points), what is it that you might want to communicate about Sprint Koala (which was three Sprints ago) to someone else during Sprint Wallaby? Other than as an abstract referent, how will its name add value to the communication?

Unless you are mapping Sprint names onto a sequence anyway, and know that Sprint Koala was the 23rd Sprint and that Sprint Wallaby is really the 26th, you can't even convey ordinal or time information without additional cognitive load. That seems like the opposite of "user-friendly" to me.

The Value of Ordinal Information

While the name of the Sprint is rarely useful, the number of Sprints in the current project plan or the team's current progress along the plan's ordinal axis do provide some utility. For example, a plan with 25 two-week sprints can be estimated to take 50 weeks. Likewise, a project that has less than 30% of the remaining Product Backlog completed after consuming 50% of the planned iterations is quite likely to be out of tolerance.

I don't know how to divide a gaggle of geese by two koalas and a lemur, so I'm pretty sure that these sorts of names provide no value in managing project metrics or even calculating simple percentages. In contrast, assigning ordinal numbers as labels provides for some useful math functions to monitor and communicate about your project, even if the contents of past and future Sprints are neither known nor tracked.

I don't know how to divide a gaggle of geese by two koalas and a lemur, so I'm pretty sure that these sorts of names provide no value in managing project metrics or even simple percentages. In contrast, assigning ordinal numbers as labels provides for some useful math functions to monitor and communicate about your project, even if the *contents* of past and future sprints are neither known nor tracked.

Incomplete Work as "Failure"

Illustrative Example

Philosophically, the question of whether an empirical control process like Scrum can "fail" at all is highly debatable. From a pragmatic business perspective, though, anything that doesn't produce a desired result is often classified as a failure.

> A core principle of the Agile Manifesto says:
>
> > Simplicity—the art of maximizing the amount of work not done—is essential.
>
> Unnecessary or premature work are wasteful activities. Lean frameworks refer to *muda*, *mura*, and *muri* to differentiate between different types of waste. Scrum works to prevent such waste through time boxing, just-enough and just-in-time planning, and by deferring implementation details until the last responsible moment.

The following question[5] is an excellent example that highlights several common anti-patterns, including:

- using task-completion rates as a proxy for productivity
- treating work-not-done as an intrinsic failure
- setting management-defined productivity targets rather than building a predictable development cadence

- imposing external process control, rather than developing a team-driven continuous improvement process

My organization's leadership has decided that if there are any user stories not fully completed per the Definition of Done at the end of the Sprint, the Sprint has failed. This "failure" also includes when user stories have been removed from the Sprint without affecting the Sprint Goal. For example:

- when the Development Team is able to meet the Sprint Goal without completing the user story; or
- when the Product Owner has agreed to pull a story out of scope because it's redundant or lower priority.

I understand the management objective here, because completing all planned work in a given Sprint sounds sensible, but there is no focus on the importance of the Sprint Goal. To me, the objective of each Sprint is to achieve the Sprint Goal. If this means a user story can to be removed without affecting the Sprint Goal, then it's not necessarily a Sprint failure.

If a selected user story was not completed in full, but the Sprint Goal was still achieved because the unfinished Story was not that vital to the Goal, is this really justification for calling this a failed Sprint? As a Scrum Master, I have a hard time with this "rule," but I'm not entirely sure how to back up my opinion.

Analysis and Advice

My organization's leadership has decided that if there are any user stories not fully completed per the Definition of Done at the end of the Sprint, the Sprint has failed.

This is not only incorrect, it's an abuse of the Scrum framework and a thorough misunderstanding of how work should be selected for inclusion into a Sprint.

The purpose of a Sprint is *not* to complete a list of user stories; that's simply a means to an end. Rather, the purpose of a Sprint is to provide a bounded time box to work on Product Backlog items that collectively deliver the value defined by an overarching Sprint Goal.

Sprint Failure Conditions

While not addressed specifically within the Scrum Guide, a Sprint has only three pragmatic failure conditions:

1. The Sprint Goal has not been met.
2. The delivered Increment is not in usable condition.
3. The Increment does not meet the "Definition of Done."

That's it. Individual stories can be done or not-done, forecasts (estimates) can be missed, and the team may have successfully delivered the wrong MacGuffin. Such Sprints are still technically "successful" in that they delivered a potentially-releasable Increment and leveraged the framework to provide the business with process transparency and appropriate opportunities to inspect-and-adapt.

Sprint Goals and Increments

The following elements of the Scrum framework are explicitly defined in the Scrum Guide. The Sprint Goal is developed during Sprint Planning, and provides guidance throughout the Sprint. The Increment is the work completed according to the Definition of Done, and is essentially the *de facto* deliverable for the Sprint.

Sprint Goal

> The selected Product Backlog items deliver one coherent function, which can be the Sprint Goal. The Sprint Goal can be any other coherence that causes the Development Team to work together rather than on separate initiatives.

Increment

> At the end of a Sprint, the new Increment must be "Done," which means it must be in useable condition and meet the Scrum Team's definition of "Done." It must be in useable condition regardless of whether the Product Owner decides to actually release it.

Educational Opportunities

In general, your management team's approach is exhibiting a number of smells that indicate a faulty Scrum implementation. It is the Scrum Master's job to educate the entire organization, including the Scrum Team and senior management, about the way Scrum actually works.

Specifically, you should use this as an opportunity to address the following project smells implied by your original post:

1. A Sprint should have a *coherent* goal.

   As defined by the Scrum Guide, each Sprint should have a defined goal which causes the team to work *together* rather than on separate initiatives. If you don't have a coherent Sprint Goal, a coherent Increment, or a collection of Sprint Backlog items that are interrelated, then the framework is being implemented incorrectly.

2. Work is *selected* by the Scrum Team, not assigned from the outside.

   The Product Owner prioritizes work on the Product Backlog. The Development Team negotiates with the Product Owner during Sprint Planning to select items from the top of the Product Backlog to be added to the Sprint Backlog. These items must:

   - Fit within the time box of a single Sprint.
   - Support the current Sprint Goal.
   - Collectively deliver a vertical slice of value.

   Assigning user stories to the Development Team, selecting unrelated Product Backlog items, dictating the contents of the Sprint Backlog to the Development Team, or a lack of active collaboration between the Product Owner and the Development Team during Sprint Planning are all huge red flags.

3. Forecasts are not money-back guarantees.

   Sprint Planning is a *short-term* planning exercise that estimates both level-of-effort and team capacity to arrive at a reasonable forecast of the work that can be completed within a single Sprint with the resources and knowledge currently available. Forecasts can be missed for a wide variety of reasons; it is *not* inherently a failure of either the framework or the team's work during the iteration. Instead, a missed forecast is an opportunity to:

   - Learn more about the problem domain.
   - Inspect-and-adapt the framework or development process.

- Improve estimating techniques.

4. It is more effective to *iterate* rather than to affix blame.

    If your organization is trying to "hold people accountable" for forecasts, they have misunderstood the difference between an estimate and a guarantee. Educate them about how iterative development and iterative process improvement actually work, and explain how the Scrum framework provides them with the tools they need to effectively manage emergent designs and processes.

Implementation Questions

This section comprises common questions about how to implement certain aspects of project management or product development within the Scrum framework.

How are Product Releases Scheduled?

Pragmatically speaking, non-agile frameworks often set their product delivery dates *first*, thereby fixing the project schedule. Scope, budget, and work packages are then defined through big, upfront planning exercised to try to shoehorn as much work into that schedule as possible.

Agile frameworks like Scrum work differently. As a rule of thumb (because there are always exceptions), Scrum trades exhaustive upfront specifications for just-in-time planning and emergent design. This paradigm shift often leads to questions like this one[6], which very reasonably asks *how* to derive a product release date without layout out a detailed project schedule.

Illustrative Example

 I've run into a constant issue. When a project starts, the client typically has a list of functionalities to be built in to the application. As a team we would like to follow Scrum, but the next thing the client asks is for a go-live date.

The client has his own market-driven deadlines. So, it's valid that he needs to provide a delivery date for the application to the rest of the business.

Since the team can't plan out more than one Sprint at a time, Scrum doesn't seem to provide any visibility on the expected end date. The client says, "I've already told you exactly what functionality I need. Tell me when you can build this app."

How do you deal with a situation like this in Scrum?

Analysis and Advice

Agile release planning in Scrum is based on *fixed-length, normed-capacity cycles* that operate on dynamically-planned and dynamically-scoped features (sometimes called "just-enough" or "just-in-time" planning). In Scrum, fixed-date release planning *must* be handled by controlling scope to meet the deadlines, as you cannot have both fixed-date and fixed-scope deadlines simultaneously. This is rarely a practical problem, but can be a political one in non-agile shops.

While not explicitly stated in the Scrum Guide or the Agile Manifesto, many experienced practitioners embrace a number of principles and practices that support iterative product development. These include:

- "Just-enough design" (sometimes referred to as YAGNI), which embodies core agile principles like simplicity, limiting non-essential work, and elimination of wasted effort. *NB: Just-enough design is often a difficult frame-shift, because it's such a huge departure from the traditional approach of big, upfront planning through detailed specifications.*
- "Just-in-time (JIT) planning," which defers design or implementation decisions until the last *responsible* moment.
- Emergent design, where the goals and implementation details of the development cycle evolve over time until the product is "good enough" for its intended purpose.

Without understanding the application of these principles, frameworks like Scrum can feel a lot like performing a highwire circus act without a safety net. Because these principles impact business functions too, successful Scrum adoption usually needs the full support of an educated senior leadership, not just participation by the members of the Scrum Team.

How to Perform Agile Release Planning

Agile release planning is based on iterations. In order to do calendar- or time-based release planning in Scrum:

1. The entire Product Backlog is given a rough estimate, usually at the level of epics rather than detailed user stories.
2. A fixed Sprint length is determined. 2-4 weeks are common values; this length should not change during the lifecycle of the project without recalculating Sprint capacity and adjusting release schedules.
3. A velocity is calculated based on the team's past performance, or on an "educated guess" if no past performance is yet available.
4. A fudge factor for the cone of uncertainty is applied. This is commonly 0.6 for new teams or for projects with large cones of uncertainty during initial project planning, but fudge factors are inherently variable and can be adjusted to fit currently-available knowledge about the problem domain, the team, and available resources.
5. A planning velocity is calculated, e.g. `estimated velocity * fudge factor`.
6. The number of iterations for a release are calculated using the following variables:
   - e = aggregate estimate of all Product Backlog items
   - v = planning velocity
   - i = estimated iterations for release, rounded up to the nearest whole number
7. The formula `e / v = i` estimates the iterations required to empty the Product Backlog.
8. The *i* value can be converted to a calendar or time estimate by multiplying interations by the length of the Sprints in weeks or months, e.g. `i * 2`.

A Worked Example

Let's say you have a total backlog of 200 story points, and plan to use a two-week Sprint length. Your team's historical velocity is 20, but this is a brand-new project with a large cone of uncertainty, so your fudge factor is the standard 0.6 multiplier; as a result, your planning velocity is 12 story points per Sprint after applying the fudge factor.

Given the planning values above, you estimate a target release release date as follows:

1. Iterations required to complete all of the current Product Backlog items is roughly 200 / 12 = 17 Sprints.
2. Time to complete the current Product Backlog is 17 * 2 = 34 weeks.

Using iteration-based calculations, your initial release plan predicts approximately 34 weeks to ship all the features currently in the Product Backlog. This is an *estimate* based on the information currently available, and should be treated as an initial planning value rather than an ironclad guarantee.

Adjust Scope During Inspect-and-Adapt Inflection Points

As the project progresses, the cone of uncertainty narrows and the team can make more accurate estimates about the amount of remaining work on the Product Backlog. In addition, a properly-functioning Scrum Team will become more accurate about measuring its velocity as the project continues, so the release-schedule calculations should be redone from time to time to "true up" the schedule based on more accurate data as it becomes available.

In addition, the Product Owner may add or remove scope (in the form of Product Backlog items) throughout the project. This will expand or reduce the scope of the project, and will obviously impact the estimated schedule. Changing project scope should generally trigger a recalculation of the release date when that happens.

Finally, Scrum strives to provide a potentially-shippable product at the end of each and every Sprint. While it may not be feature-complete in the sense that it contains 100% of all Product Backlog items, the product should be in a stable and releasable state during each Sprint Review. The organization can choose to ship earlier than planned if sufficient value is present in the product to justify shipping it in its current state. This "cashing out" of earned value to ship a viable product that is deemed "good enough" can provide *the business* (not just the Scrum Team) with a significant agile advantage.

Why Define a Sprint Goal Each Sprint?

Illustrative Example

The Sprint Goal is one of the most essential (but often overlooked) aspects of Sprint Planning. Whether a Sprint is successful or not isn't based on how much work is done or effort expended, but rather on whether the Scrum Team has successfully met the current Sprint Goal.

As exemplified in the following question[7], many teams ask why they need a Sprint Goal other than "complete all the items in the Sprint Backlog." Understanding the *value* of the Sprint Goal is critical to making the most of the framework.

As a Scrum Team, we are committed to deliver Product Backlog items per the Product Owner's priority. The team doesn't understand why we should define Sprint Goals. How does having a Sprint Goal for each Sprint help?

Analysis and Advice

You must define a Sprint Goal for each Sprint for two primary reasons:

1. It's *required* by the framework for both philosophical and pragmatic reasons as described in the rest of this chapter.

2. It provides focus for the Scrum Team's limited resources, including the finite timeboxes used in iterative frameworks like Scrum.

While it is *possible* to be agile without a central coherence as provided by the Sprint Goal, the result isn't Scrum. Furthermore, Scrum (or other coherence-based methodologies) may not be the right fit for your organization or project if Sprint Goals can't be constructed or honored.

The End Note of the Scrum Guide clearly states (emphasis added):

> Scrum's roles, events, artifacts, and rules are immutable and **although implementing only parts of Scrum is possible, the result is not Scrum**. Scrum exists only in its entirety and functions well as a container for other techniques, methodologies, and practices.

Scrum is a framework. You can modify certain aspects of its implementation, or combine it with industry best practices like Kanban boards, pair programming, or continuous deployment, but you can't ignore core requirements like defining a Sprint Goal each Sprint. If you do, you're practicing ScrumBut, *not* implementing Scrum. This generally leads to blaming the Scrum framework rather than its misapplication if the project doesn't run smoothly.

Why the Sprint Goal is Needed

The Sprint Goal is an essential element of Scrum, which is primarily (although not exclusively) a *product development* framework. As such, it is intended to iteratively deliver features, as opposed to other frameworks which focus on other mechanics such as throughput or cycle time. Because of this development-centric focus, work is intended to be batched into cohesive themes for each iteration.

If you are "doing Scrum" without setting well-defined Sprint Goals, then you can't actually claim to be following the Scrum model. This isn't just a philosophical distinction. The Sprint Goal is essential to a properly-implemented Scrum framework. The Scrum Guide defines the Sprint Goal as follows:

> The Sprint Goal is an objective set for the Sprint that can be met through the implementation of Product Backlog. It provides guidance to the Development Team on why it is building the Increment. It is created during the Sprint Planning meeting. The Sprint Goal gives the Development Team some flexibility regarding the functionality implemented within the Sprint. The selected Product Backlog items deliver one coherent function, which can be the Sprint Goal. The Sprint Goal can be any other coherence that causes the Development Team to work together rather than on separate initiatives.
>
> As the Development Team works, it keeps the Sprint Goal in mind. In order to satisfy the Sprint Goal, it implements functionality and technology. If the work turns out to be different than the Development Team expected, they collaborate with the Product Owner to negotiate the scope of Sprint Backlog within the Sprint.

In other words, the Scrum Goal provides:

1. A measurable objective for the time box.
2. Context (and therefore scope) for the Product Backlog items developed within the time box.

3. Context for the daily Scrum, without which it typically devolves into status reporting.
4. A baseline expectation for measuring variance.
5. The essential criterion for cancelling a Sprint due to superseded business goals, adaptive learning, or as an escape hatch for errors in Sprint Planning or delivery.

In my professional experience, without the Sprint Goal as a core element, Scrum often fails to provide expected gains in productivity.

Reasons People Skip Sprint Goals

Teams that don't implement Scrum Goals often do this for some common reasons, including:

1. **They aren't really doing iterative time boxing.**

    If you aren't respecting the time box, or the team is doing demand-based work rather than building features within a time box, then having an over-arching goal probably doesn't make sense. However, that's often a "project smell" that Scrum is the wrong framework for managing the project.

2. **The team, and especially the Product Owner, isn't leveraging the framework.**

    Scrum is meant to deliver a product in thin, preferably-vertical slices of functionality. However, there's a lot of effort that goes into prioritizing Product Backlog into features and related stories that fit within a single time box. It's hard, so skipping the Sprint Goal is essentially a cop-out.

3. **The 100% utilization fallacy and lack of prioritization.**

    Some teams, when under pressure to deliver more work in less time, discard the Sprint Goal in order to work on multiple objectives at the same time or to avoid slack in the process. However, deliberate disconnection of a central coherence and multiple priorities competing for limited resources are sure-fire ways to torpedo any possible efficiencies from an iterative methodology.

I'm sure there are other reasons, too. The point here is that, unless you're using Scrum for something other than product development (e.g. administrative or service-oriented programs), failing to define *and defend* a central coherence is not only canonically-wrong from a framework perspective, but frequently detrimental to the project's allocation of focused resources. This notion of *focused (but limited) resources* is absolutely essential to the iterative paradigm, and underlies virtually all agile and traditional methodologies. Scrum just makes it much more explicit through the use of the Sprint Goal.

What Next?

If a team is struggling to define coherent Sprint Goals for each iteration, the Scrum Master can and should educate the team (and the organization!) on the importance of setting and adhering to the goals. This should be done not only because it's important to the Scrum framework, but because it is ultimately important to the product development effort.

If the Scrum Master isn't able to provide this level of guidance, or doesn't have sufficient influence within the organization to enforce the framework's essential components, then hiring an Agile Coach to advise the team would be the logical next step. The coach can help the team improve the implementation of Scrum, or suggest alternative frameworks that can succeed without an explicit central coherence for each time box.

Who Manages Project Budgets?

Illustrative Example

When adopting Scrum, one of the most confounding questions for many organizations is how to integrate their traditional budgeting process with the notion of a "self-organizing" Scrum Team. It sounds like a recipe for anarchy, or at least reckless spending!

The following question[8] is so important to the business, including executive leadership and line management, that it probably deserves its own book.

 I have read many books about Scrum, including blogs and academic articles. All of them dodge core business questions like:

- Who is responsible for managing the project's budget?
- Why is this unnamed person or role responsible, versus someone else?

Does the Scrum framework address these important questions in any way?

Analysis and Advice

From a purely pragmatic perspective, the Product Owner is responsible for directing the *allocation* of the project's budget, while the funding source of the budget varies by organization.

Funding a Project

Projects are funded in organization-specific ways. Usually, a project sponsor gets funding from a steering committee or the Chief Financial Officer (CFO). In other cases, a project may be funded out of a specific department's capital or operating budgets. In all cases, there is ultimately a budgetary authority that authorizes predefined or ongoing expenditures for a project.

Any well-run organization has a budgeting process. This process delegates formal authority to someone in the organization to manage that budget. On projects using the Scrum framework, that person *should* be the Product Owner.

Managing Resources via Product Backlog

In Scrum, the Product Owner is responsible for prioritizing the deliverables. By extension, this makes the Product Owner responsible for the allocation of resources for the project. Let's look at a concrete example of how this works.

On a recent project, there were two user stories on the Product Backlog. The first story was a customer-facing feature, while the second was an infrastructure story for building a continuous integration (CI) server. Both stories represented a cost to the project, but the CI story required more upfront expenditures (i.e. server equipment, colocation fees, and ongoing maintenance costs) to be charged against the project's budget.

Prioritization of Product Backlog items therefore encompasses budget management, both implicitly and explicitly. In this case, the Product Owner needs to decide between capital expenditures and operational costs. Should she spend part of the budget to buy new CI servers for the project to help automate testing, or should she allocate more of the finite budget towards labor costs for manual testing? She has to make choices.

If she de-prioritizes the CI story, the project may need more manual testing resources, or perhaps fewer stories will be accepted into each Sprint by the Development Team because of limited testing capacity. Higher labor requirements can impact scheduling and long-term staffing costs for the project, but might be acceptable trade-offs in exchange for being able to ship revenue-generating features at the end of a near-term Sprint.

In this particular case, the Product Owner chose to prioritize the feature story and de-prioritize the CI story. **In effect, the PO managed the budget *through* the Product Backlog.** This is the primary vehicle the Product Owner has for managing the project's budget.

Additional Resources

As the question correctly pointed out, it's hard to find a concrete reference to this budgetary responsibility, perhaps because most practice guides focus on the Scrum Master or Development Team roles in Scrum. For more on the Product Owner role, including at least 12 specific references to the Product Owner's responsibility for budgets, see *Agile Product Management with Scrum** by Roman Pichler.

**Agile Product Management with Scrum: Creating Products that Customers Love.* Pichler, Roman. Addison-Wesley, 2010.

Must Sprint Backlogs Be Completed?

Illustrative Example

During Sprint Planning, prioritized Product Backlog items that fit the current Sprint Goal are selected by the Development Team (in collaboration with the Product Owner) for inclusion in the Sprint Backlog. The Sprint Backlog is an artifact that belongs to the Development Team; it serves as the team's planning guide for the current Sprint. It's neither a fixed specification nor an immutable work breakdown structure.

As the following question[9] shows, this process often creates confusion about what the Development Team has forecast it can deliver by the end of the Sprint.

I'm working in development team working on an online mobile game. We've been doing Scrum for several Sprints now.

There are different reasons each Sprint, but we never seem to be able to finish all the items selected for the Sprint. Is this normal? If we can't finish all the tasks in our Sprint Backlog, does this mean we are doing Scrum wrong?

Analysis and Advice

The failure here is *not* that the Development Team isn't completing all its work items; the failure is that the Scrum Team lacks a central coherence for each Sprint. In other words, you need a Sprint Goal as a central feature of Sprint Planning.

 In Scrum, a "central coherence"[10] doesn't just represent the theme of the current Sprint, or even the aggregate work product representing the Increment to be delivered at the end of the Sprint. A coherence represents a collective effort by the whole team. It can be *anything* that requires the Development Team to work together collaboratively, rather than working individually or on non-intersecting tasks.

 Always remember that the goal of a Sprint isn't to complete lots of backlog items. **The goal of a Sprint is to *accomplish the Sprint Goal*.**

A focus on "doing all the things" is an anti-pattern that stems from non-agile assumptions about how work should be divvied up and performed. It's also very commonly a form of the *100% utilization fallacy* that attempts to maximize the utilization of individual team members, rather than optimizing for the throughput of potentially-shippable features that meet a complete Definition of Done.

Because the team is being asked to work on many disparate tasks, rather than collaborating in a cross-functional way on completing vertical slices of work, the results you're seeing are almost entirely predictable. You need to shift your focus away from pseudo-productivity in the form of task-completion towards a focus on *feature-completion* to resolve this problem.

In other words:

1. Ensure each Sprint Planning session produces a coherent Sprint Goal.
2. Ensure the work selected for the current Sprint aligns with the Sprint Goal.
3. Ensure everyone on the team is working *together* on the Sprint Goal.
4. Measure the success of the Sprint on whether or not the Scrum Team was able to meet the Sprint Goal.

Doing anything else is not Scrum. Doing anything else will also be painful, frustrating, and demoralizing. So, unless you are an organization of professional masochists, stop doing what you're doing and implement Sprint Goals as a core practice.

 Ultimately, whether you complete all of the items on a given Sprint Backlog or not is immaterial. What matters is whether or not the Scrum Team collectively met the agreed-upon Sprint Goal for the current Sprint.

What is Scrum's Process Overhead?

Illustrative Example

All project management frameworks (even the agile ones!) have some inherent process overhead. Scrum is often touted as a lightweight framework, but is that actually true? As the following question points out, the answer to this question isn't easily found in most Scrum-related materials.

 Each of the three Scrum roles (Product Owner, Scrum Master, and Development Team member) is full-time job when performed properly. However, some of that time is inevitably spent on framework overhead rather than directly on product development. This is true regardless of framework, because *all* frameworks require some amount of process overhead. So, how much process overhead does Scrum actually entail?

Analysis and Advice

All frameworks entail some amount of process overhead. In Scrum, some of that overhead is in the form of hours worked by the Product Owner and Scrum Master. While they are essential to the framework process, some might consider their labor hours as "overhead" since they aren't directly tied to hands-on product development activities. Other overhead is a byproduct of implementing Scrum's controls and artifacts.

While Scrum is often described as a lightweight process, it actually carries significant process overhead that *increases as cycle time contracts*. Shorter Sprints have faster feedback loops and more inspect-and-adapt opportunities, but have more overhead and reduced development time per iteration. Longer Sprints have slower feedback loops and longer cycle times, but offer reduced overhead and increased development time per iteration. Finding the right set of trade-offs for a given project is an important part of successful Scrum adoption.

 Longer Sprints increase cycle times but reduce process overhead. Other variables such as team size, meeting length, and process maturity will have an impact, too. These calculations also exclude participation in other organizational activities such as Scrum-of-Scrums meetings, stakeholder meetings, team "hallway meetings," and other daily tasks that are essential to the product development process. So, consider the whole thing a reasonable initial planning value for predicting overhead, rather than a theoretical maximum.

Calculating Overhead

A typical labor estimate for an 8-person Scrum Team using two-week Sprints might look like this:

Sprint Planning
    4 hours per person each Sprint
    32 hours per Scrum Team each Sprint

Daily Scrum
>15 minutes per person each day

>20 hours per Scrum Team each Sprint

Reporting and Artifact Management (e.g. burn-down charts)
>1 hour per Scrum Master per day

>10+ hours per Scrum Master per Sprint

Backlog Refinement
>4 hours per person per Sprint

>16 hours per Scrum Team per Sprint

Sprint Review
>2 hours per person each Sprint

>16 hours per Scrum Team each Sprint

Sprint Retrospective
>2 hours per person each Sprint

>16 hours per Scrum Team each Sprint

Total Overhead
>87 hours for the Development Team

>23 hours for the Scrum Master and Product Owner

>110 hours for the Scrum Team as a whole

> The Scrum Guide states that refinement should consume no more than 10% of the Development Team's capacity—that's 6-8 hours per person, or upwards of 36 hours per team, in a two-week Sprint!—but a mature Scrum Team often needs much less. To reduce the overhead of this particular framework event, the Product Owner and Scrum Master can ensure that Product Backlog are as well-refined and estimable as possible before involving the Development Team.

An 8-person Scrum Team has a standard run-rate of 320 person-hours for each two-week Sprint. With 87 hours of overhead for the Development Team, and an additional 23 hours collectively allocated against the Scrum Master and Product Owner, more than 34% of each two-week Sprint is "lost" to process overhead!

> Please note that all *healthy* projects, regardless of the project management framework, need to allocate similar blocks of time for attending meetings, collecting/disseminating status, and generating framework artifacts. Scrum simply makes this overhead visible.
>
> Some methodologies (like Scrum) trade time spent in face-to-face events for less time spent on reporting or artifact management. Other methodologies exchange fewer team meetings for more project manager and line management overhead, such as analyzing additional metrics or generating more reporting artifacts.

Don't Panic!

None of this is to say that Scrum carries an unacceptable level of overhead. Scrum values transparency, and simply makes its overhead visible as a cost to the project rather than sweeping it under the rug or wishing it into the cornfield. In addition, much of this overhead:

1. directly feeds collaborative communications within the team and with key stakeholders;
2. drives just-enough and just-in-time planning; and
3. provides effective feedback loops that improve quality, fitness-for-purpose, and limit sunk costs.

When you consider the oft-cited statistic from AIG Consulting that 68% of IT projects fail, any costs associated with this framework overhead are negligible when compared to the sunk costs that frequently come from failed projects with extremely long lead times. By shortening the project feedback cycle to 30 days or less, Scrum enables executive leadership to make more inspect-and-adapt decisions for the business, including shipping products that are "good enough" *right now* in order to cash out earned value from the project at the end of any iteration, or terminating failing projects before they accrue significant sunk costs.

If you decide to adopt Scrum, don't do it because it's "lightweight" or "low overhead." Do it because it adds value to your product development process, improves product quality, or reduces sunk costs. In the end, the business case for adopting a fully-transparent framework with clearly-defined process controls and easily-calculable tradeoffs practically writes itself.

Backmatter

About the Author

May 9, 2017

Todd A. Jacobs is the Chief Technology Officer and resident agile guru at CodeGnome Consulting, LTD. He is also a thought leader on Scrum adoption and DevOps transformations at Project Management Stack Exchange.

Intensely passionate about technology and process, Todd has published numerous technical articles for magazines such as *SysAdmin*

Magazine and *Linux Journal*. He also participates actively on Stack Overflow, especially in the Ruby, Linux, and various shell scripting categories.

Most importantly, Todd enjoys engaging with his readers, so please reach out through this book's:

1. reader forum, where you can ask questions about the book or suggest improvements for future versions; or
2. feedback page about anything else.

Thank you for reading this book. Please do stay in touch!

Notes

Anti-Patterns

Daily Scrum as Status Pulls

1 Question adapted from a post by Prisoner ZERO, under a Creative Commons Attribution-ShareAlike 3.0 Unported license.

Sequential Development and Testing

2 Question adapted from a post by John, under a Creative Commons Attribution-ShareAlike 3.0 Unported license.

Expecting Ever-Increasing Velocity

3 Question adapted from a post by Pedro, under a Creative Commons Attribution-ShareAlike 3.0 Unported license.

Sprint Names as "Permalinks"

4 Question adapted from a post by Bijay Rungta, under a Creative Commons Attribution-ShareAlike 3.0 Unported license.

Incomplete Work as "Failure"

5 Question adapted from a post by Dave, under a Creative Commons Attribution-ShareAlike 3.0 Unported license.

Implementation Questions

How are Product Releases Scheduled?

6 Question adapted from a post by John, under a Creative Commons Attribution-ShareAlike 3.0 Unported license.

Why Define a Sprint Goal Each Sprint?

7 Question adapted from a post by ssharma, under a Creative Commons Attribution-ShareAlike 3.0 Unported license.

Who Manages Project Budgets?

8 Question adapted from a post by Pomario, under a Creative Commons Attribution-ShareAlike 3.0 Unported license.

Must Sprint Backlogs Be Completed?

9 Question adapted from a post by Paiman Roointan, under a Creative Commons Attribution-ShareAlike 4.0 International license.

10 See "Sprint Goal" in the Scrum Guide's section on Sprint Planning for a more complete definition.

www.ingramcontent.com/pod-product-compliance
Lightning Source LLC
Chambersburg PA
CBHW070442220526
45466CB00004B/1755